This is book is dedicated to all of you up and coming electricians that are busting your asses every day on the job, studying to get further, researching, asking questions, watching videos, and engaging in forums. You are driven to excel at a lifelong craft that will provide a better life for you and your family. Keep doing good work my dude, and you'll get there...just keep grinding.

-DS

Understanding The NEC Layout - Part I <u>Chapters</u>

One thing that helps you understand the NEC is taking a birds-eye-view of how it's laid out as a manual. The Chapters are not just randomly laid out as you may think - there's a purpose in HOW it is laid out and why.

The 2017 NEC has a table of contents, 9 Chapters, 10 informational annexes, and an index - all of which serve a certain topic or purpose. Knowing the layout of these Chapters is crucial in being able to work through the book to find codes both on the job as well as in the testing center, while taking an exam. Let's take a look at it from above:

Table of Contents

Chapter 1 – General
Chapter 2 – Wiring and Protection
Chapter 3 – Wiring Methods and Materials
Chapter 4 – Equipment for General Use
Chapter 5 – Special Occupancies
Chapter 6 – Special Equipment
Chapter 7 – Special Conditions
Chapter 8 – Communications Systems
Chapter 9 – Tables
Annex A – Product Safety Standards
Annex B – Application Information for Ampacity Calculation
Annex C – Conduit and Tubing Fill Tables for Conductors and Fixture Wires of the Same Size
Annex D – Examples
Annex E – Types of Construction
Annex F – Availability and Reliability for COPS and FPTs forCOPS
Annex G – Supervisory Control and Data Acquisition (SCADA)
Annex H Administration and Enforcement
Annex I – Recommended Tightening Torque Tables from UL Standard 468A-B
Annex J – ADA Standards for Accessible Design
Index

Table of Contents and Index
All of that may seem overwhelming, but most of the Annexes are not going to be on the test – in most states. I've color-coded the above information because each item that is similarly colored can be thought of as having a combined purpose. For example, the table of contents and index are both just tools for navigating topics and keywords in the book. They're your maps essentially - they hold all of the key destinations and locations you're looking for in the book.

Chapters 1-4
Notice that Chapters 1-4 are combined into one conceptual group. These Chapters are rules that generally apply to all places and equipment you're going to run into. This is the bedrock/foundation of all of the code knowledge in the NEC and what you'll rely on in most cases. These Chapters are standards for homes, apartments, banks, strip malls - normal, everyday places and situations that most electricians are going to come across on a daily basis.

Chapters 5-7
Chapters 5-7 are all grouped together as well, because these 4 Chapters are "special cases" (locations, equipment, situations) that supplement or modify information in Chapters 1-7. If you're in an environment where a lot of people assemble (like a concert), or in a corrosive environment (around large bodies of water), you'll look in Chapters 5-7. Not just the location, but the equipment can be specialized as well. If you're working on equipment that is not standard air conditioners and water-heaters, such as cranes, welders, elevators, and x-ray equipment, you'll look to Chapters 5-7 for these rules.

Chapters 8-9 and Annex C
The reason I have Chapters 8, 9, and Annex C grouped together is that, in most states, you may not be tested on these, but they may help you find information that can get you answers to test questions. They have similar information and functions. Chapter 8 applies to communications circuits, most of which apply to low voltage systems and are not required code. In Article 90.3, the NEC states, "Chapter 8 covers communication systems and is not subject to the requirements of Chapters 1-7, except where the requirements are specifically reference in Chapter 8."

Chapter 9 is used mainly for conductor and conduit dimensions, and being able to figure out how many conductors can fit into what sized conduit. More specifically if you're trying to fit multiple differently sized conductors into a run of conduit. Annex C holds table after table of different conduits, sizes, and how many of the same sized conductors you can fit in whatever conduit you're working with. So Chapter 9 – different sizes. Annex C – same sizes. Keep in mind the values in the Annex C tables are based on the information in the Chapter 9 tables. Annex C is really a quick reference for us out in the field, because more often we use the same sized conductors in conduit, instead of differently sized conductors.

Now when you're looking for codes about installing receptacles near a pool, you know you can save time by not looking in Chapters 1-4. It has to be somewhere in Chapters 5-7. You also know it's not in any of the Annexes. Once you dig around and start getting a feel for which Articles are in which Chapters, it really helps you understand how to navigate the NEC and quickly to find information.

> ***PRO TIP – *Get yourself a set of colored NEC approved tabs for your code book. These are allowed, in most states, to be installed in your code book so that you can quickly identify Chapters, tables, articles, and sections that are specific and used frequently.* ***

means that this wording was changed or added from a previous edition of the code. There are also small, bold, black dots (•) that appear to the left side of the paragraphs throughout the code. They are positioned cleverly in between paragraphs and each one of these dots represents content that was removed from the current or prior codes. There is some numbering that stays relatively consistent throughout the book. Most of the "scope" sections are numbered similarly – for example 312.1, 322.1, 620.1, etc. all give you the main scope of the article at hand. Similarly 605.2, 504.2, and 800.2 are all common because they have definitions that are not in the beginning of the NEC in Article 100. That's right, rather than putting all of the definitions for everything in the NEC in one place (Article 100), the writers thought it would be fun to have definitions appear all over the damn place throughout the book. But hey, at least they gave you a .2 identifier!

Definitions

The last stop on the Crazy Train (insert Ozzy screaming and Randy Rhoads ripping a tasty guitar lick here) is the Definitions. Article 100 (Chapter 1) is dedicated to Definitions. This is something I would study as much as you can. Knowing terms like "utilization equipment", "grounding vs grounded, vs ungrounded, vs bonding", and "continuous duty" can be crucial in your understanding of the NEC code. This stuff was not written by electricians, it was written by a lot of guys that wear slacks and ties and get paid a lot more than we do. It doesn't have to follow logical sense, it has to protect lives and assets. There are Definitions in a lot of Articles, but not all of them. They only appear in Articles that introduce new equipment, locations, or circumstances. The Definitions in the front of the book (Article 100) are more relative to the entire code book which is probably why they separated them like they did.

Understanding The NEC Layout - Part II Articles, Parts, Sections, and Sub-Sections

The next thing you need to wrap your head around is the meat and cheese of what's inside these Chapters. This information is also laid out purposefully, though it'll seem like an unnavigable maze at times.

Articles

Articles are the main structure of each Chapter. For example, in Chapter 3 we have Articles 300, 310, 312, 314, and so on all the way to 399. There doesn't seem to be any recognizable convention as to why certain numbers are skipped or chosen, other than possibly leaving space for future code cycles. These articles are what you'll need to pay a lot of attention to, as they are the more like sub-Chapters to each Chapter. There are a lot of articles in the NEC, and some Chapters have a ridiculous amount of them. The information can be a bit confusing, so the writers of the NEC came up with Parts.

Parts

Parts are in every Article, because a lot of topics differ wildly and need to be grouped together in a way that makes navigating the Article easier. We have Article 250, which covers all of grounding and bonding for the entire code book. Inside of it, there is so much information that there are 10 parts. These are roman numerals - i.e. Part I, II, III, IV. These Parts let you know that there are very different topics in the Article - in this case Grounding and Bonding. Part I is general grounding and bonding, whereas Part VIII is direct current systems and Part V is bonding and how it relates to several other Parts. All of these Parts correlate to the main Article 250 but each is a separate subject on its own. There is a great diagram in the 2017 NEC on page 70-104 Figure 250.1 that shows the breakup of the 10 Parts in Article 250 and how each differs.

Sections

This is where things start to get a bit hairy. Within each Article, we also have Sections. Each Section is a more specific situation that deals with the main Article. Some Articles have many sections, while others have very few. Some Examples of Sections are:
- 250.50 Grounding Electrode System
- 250.78 Common Grounding Electrode
- 250.97 Bonding for Over 250 Volts

In the above examples, we're in Chapter 2, Article 250, Section 50, 78, and 97 respectively. But it gets deeper yet my friends.

Sub-Sections

A Sub-Section may be full of specific rules that have to be followed, or a list of possible rules in which only one or two need to be observed. An example of this is in
314.28. In this Section we have 314.28(A) through (E). 314.28 deals with Pull and Junction Boxes and Conduit Bodies, however there are specific situations that need to be observed. For example, are the conductors coming straight through the box, entering and exiting at 90 degree angles, are there splices? So to address these different situations, they've added a Sub-Section for each of them and labeled them (A), (B), (C), (D), and (E).

Finally we arrive at Sub-Sub-Sections. This is really more of a term that I use comically, but it's to point out that each sub-section can have sub-sections and sub- sub-sections... which really means that it is possible to have a sub-sub-sub-section. Maybe there's a better way of identifying them, but you get my point. Some examples of this would be:

314.28(E)(1) - Installation. This is sub-sub-section (1) of sub-section (E), in Section 28, of Article 314, in Chapter 3. Are you about to rip your hair out yet?! Ya.... welcome to becoming an electrician. But we're not out of the rabbit hole yet my friend. There are places like 250.52(A)(3)(1) that deal with Concrete-Encased Electrodes, and rules/options for which method you'd like to use. And even crazier yet is 310.15(B)(3)(4)(a) that deals with adjustment factors to apply to conductors depending on where, how many, and what type of conductor is being used. Ya dude... really.

Exceptions and Informational Notes

Exceptions and Informational Notes are getting grouped together because you're rarely going to use either of them. You may use Exceptions as they excuse you from following code from time to time, by making allowances that mean the code is unnecessary or cannot be followed because of the environment/equipment you're dealing with. Informational Notes are different, they're simply friendly little tips that give you extra information and references that may help you make a better decision when stumped about something. Exceptions can be, but Information Notes and Annexes are not testable in most states.

Other "Stuff" To Know

I do believe that is it. Well it's really not, there are small nuances to the book like the letter "N" appearing to the left of a paragraph, or seeing gray hi-lighting appear in random places. But those are only there to let you know that N means this whole paragraph is newly added and is not in older editions of the NEC, and hi-lighting means that this wording was changed or added from a previous edition of the code. There are also small, bold, black dots (•) that appear to the left side of the paragraphs throughout the code. They are positioned cleverly in between paragraphs and each one of these dots represents content that was removed from the current or prior codes. There is some numbering that stays relatively consistent throughout the book. Most of the "scope" sections are numbered similarly – for example 312.1, 322.1, 620.1, etc. all give you the main scope of the article at hand. Similarly 605.2, 504.2, and 800.2 are all common because they have definitions that are not in the beginning of the NEC in Article 100. That's right, rather than putting all of the definitions for everything in the NEC in one place (Article 100), the writers thought it would be fun to have definitions appear all over the damn place throughout the book. But hey, at least they gave you a .2 identifier!

Definitions

The last stop on the Crazy Train (insert Ozzy screaming and Randy Rhoads ripping a tasty guitar lick here) is the Definitions. Article 100 (Chapter 1) is dedicated to Definitions. This is something I would study as much as you can. Knowing terms like "utilization equipment", "grounding vs grounded, vs ungrounded, vs bonding", and "continuous duty" can be crucial in your understanding of the NEC code. This stuff was not written by electricians, it was written by a lot of guys that wear slacks and ties and get paid a lot more than we do. It doesn't have to follow logical sense, it has to protect lives and assets. There are Definitions in a lot of Articles, but not all of them. They only appear in Articles that introduce new equipment, locations, or circumstances. The Definitions in the front of the book (Article 100) are more relative to the entire code book which is probably why they separated them like they did.

PRACTICE TEST

100 Practice Questions For The Residential Electrician

Based on the 2017 NFPA 70 National Electrical Code

1) An insulated grounded conductor of 4 AWG or larger shall be identified by which one of the following means:

   a) A continuous black outer finish
   b) A continuous white outer finish
   c) Three continuous green stripes
   d) None of the above

2) Overhead conductors for festoon lighting shall not be smaller than 12 AWG unless the conductors are:

   a) Conductors that are listed use in damp locations
   b) Of the type THWN, THHN, or XHHW
   c) Supported by messenger wire
   d) No longer than 50ft in length

3) Where connected to a branch circuit supplying two or more receptacles or outlets, a 30- ampere receptacle shall not supply a total cord-and-plug-connected load in excess of:

   a) 16-amperes
   b) 25-amperes
   c) 30-amperes
   d) 24-amperes

4) A conductor installed on the supply side of a service or within a service equipment enclosure(s), or for a separately derived system, that ensures the required electrical conductivity between metal parts required to be electrically connected is a:

   a) Supply-Side Grounding Conductor
   b) Bonding Conductor
   c) Supply-Side Bonding Jumper
   d) Grounding Electrode Conductor

5) In a(n)_____ system, electrical equipment, wiring, and other electrically conductive material likely to become energized shall be installed in a manner that creates a low-impedance circuit from any point on the wiring system to the electrical supply source to facilitate the operation of overcurrent devices should a second ground fault from a different phase occur on the wiring system.

   a) Ungrounded
   b) Grounded
   c) 3-phase 4-wire
   d) 1-phase 3-wire

6) The earth shall be considered as an effective fault-current path.

   a) True
   b) False

7) The branch-circuit rating for an appliance that is a continuous load, other than a motor- operated appliance, shall not be less than_____ of the marked rating, or not less than _____ of the marked rating if the branch-circuit device and its assembly are listed for continuous loading at 100% of its rating.

   a) 125%, 100%
   b) 100%, 125%
   c) 83%, 100%
   d) 100%, 83%

8) Exposed, normally non-current-carrying metal parts of fixed equipment supplied by or enclosing conductors or components that are likely to become energized shall be connected to an equipment grounding conductor under which of the following conditions:

   a) Where equipment operates with any terminal at over 150-volts to ground
   b) Where supplied by a wiring method that provides an ungrounded conductor for short sections of metal enclosures
   c) Where within 9ft horizontally of ground or grounded metal objects
   d) Where located in an isolated wet or damp location

9) All pull boxes, junction boxes, and conduit bodies shall be provided with covers compatible with the box or conduit body construction and suitable for the conditions of use. Where used, metal covers shall:

   a) be oversized 3/8in to allow for expansion
   b) be used on non-metallic conduit bodies of 2in or larger
   c) comply with the grounding requirements of 250.110
   d) be listed for use in wet environments

10) For cord-connected equipment such as room air conditioners, household refrigeration and freezers, drinking water coolers, and beverage dispensers, a separable connector or a(n) _____ shall be permitted to serve as the disconnecting means.

   a) none of these
   b) toggle switch
   c) weatherproof cord cap
   d) attachment plug and receptacle

11) Receptacles shall be mounted in identified boxes or assemblies. The boxes or assemblies shall be securely fastened in place unless otherwise permitted elsewhere in this code. Screws for the purpose of attaching receptacles to a box shall be of the type provided with a listed receptacle, or shall be machine screws having_____threads per inch:

   a) 18
   b) 32
   c) 21
   d) 20

12) Receptacle outlets in or on floors shall not be counted as part of the required number of receptacle outlets unless located within how many inches of the wall?

   a) 6 inches
   b) 12 inches
   c) 18 inches
   d) 24 inches

13) A receptacle outlet shall be installed at each countertop and work surface that is_____or wider.

   a) 6 inches
   b) 12 inches
   c) 18 inches
   d) 36 inches

14) At least_____receptacle(s) outlet shall be installed in bathrooms within_____feet of the outside edge of each basin.

   a) 2, 3
   b) 1, 2
   c) 1, 6
   d) 1, 3

15) A single-phase 3-wire 200-amp service is constructed at a residence. What size copper grounding electrode conductor needs to be installed on this service?

   a) #4 copper
   b) #2 copper
   c) #6 copper
   d) #8 copper

16) When installing conductors in a rigid metal conduit that is buried under 3 inches of concrete, to what depth does this conduit need to be buried?

   a) 12 inches
   b) 6 inches
   c) 18 inches
   d) 24 inches

17) Raceways shall be used as a means of support for other raceways under which condition?

   a) Where the raceway or means of support is identified as a means of support
   b) Where the raceway is used to support boxes or conduit bodies in accordance with 314.23
   c) Both of these
   d) Neither of these

18) Switches or circuit breakers shall not disconnect the grounded conductor of a circuit.

   a) True
   b) False

19) Snap switches rated 20-amperes or less directly connected to aluminum conductors shall be listed and marked_____.

   a) ALM/CU
   b) CO/ALR
   c) For use in wet environments
   d) Aluminum conductors only

20) Fixed electric space heating equipment and motors shall be considered what type of load?

   a) Inductive
   b) Intermittent
   c) Non-continuous
   d) Continuous

21) In a dwelling unit, which of these locations are NOT required to be protected by and arc- fault circuit interrupter?

   a) Bedrooms
   b) Garages
   c) Laundry areas
   d) Hallways

22) GFCI protection must be provided for outlets that supply dishwashers installed in dwelling unit locations.

   a) True
   b) False

23) In a bathroom where receptacles are installed within 6 feet from the____edge of the bowl of a sink, they must be GFCI protected.

   a) Bottom outside
   b) Top outside
   c) Point drawn horizontally at the center of the basin
   d) Top inside

24) A 1/0 copper grounding electrode conductor is appropriate for what size ungrounded service entrance conductors?

   a) 2/0 copper
   b) 3/0 copper
   c) Over 350 – 600 copper
   d) Over 3/0 – 350 copper

25) No parts of cord connected luminaires, chain, cable, or cord suspended luminaires, lighting track, pendants, or ceiling suspended (paddle) fans shall be located within a zone measuring__horizontally and_____vertically from the top of the bathtub rim or shower stall threshold.

   a) 3 feet, 8 feet
   b) 4 feet, 8 feet
   c) 2 feet, 6 feet
   d) 3 feet, 6 feet

26) Any insulated cable 0-750-volts to ground, supported on and cabled together with a solidly grounded bare messenger or solidly grounded neutral conductor must maintain a clearance of____feet in any direction from the water level, edge of water surface, base of diving platform, or permanently anchored raft.

   a) 6.9
   b) 18
   c) 25
   d) 22.5

27) The ampacity of the branch circuit conductors and the rating or setting of over current protective devices shall not be less than_____of the total nameplate rated load.

   a) 100%
   b) 83%
   c) 125%
   d) 115%

28) Temporary electric power and lighting installations shall be permitted for a period not to exceed_____ for holiday decorative lighting and similar purposes.

   a) 30 days
   b) 90 days
   c) 60 days
   d) 120 days

29) On a 4-wire, delta connected system where the midpoint of 1 phase winding is grounded, only the conductor or busbar having the higher phase voltage to ground shall be durably and permanently marked by an outer finish that is_____ in color, or by other effective means.

   a) Yellow
   b) Orange
   c) Purple
   d) White

30) Exposed live parts on one side of the working space and grounded parts on the other side of the working space must have_____ of clearance in a 277-volt to ground environment.

   a) 3 feet 6 inches
   b) 3 feet
   c) 4 feet
   d) 2 feet 6 inches

31) Which outdoor enclosure types are all approved for wind-blown dust applications?

   a) 3, 3s, 3rx
   b) 3r, 3s, 3rx
   c) 3sx, 3r
   d) 3s, 3x, 3sx

32) For devices with screw shells, the terminal for the____ conductor shall be connected to the screw shell.

   a) Grounded
   b) Ungrounded
   c) Grounding
   d) Equipment grounding

33) A conducting object through which a direct connection to earth is established is termed a(n)____.

   a) Equipment grounding electrode conductor
   b) Grounded conductor
   c) Grounding electrode
   d) Ground bus bar

34) It is appropriate to use the "optional feeder and service load calculation" method to a dwelling unit having the total connected load served by a_____ or_____ set of 3 wire service or feeder conductors with an ampacity of 100-amps or greater.

   a) 120/240-volt, 208Y/120-volt
   b) 120/240-volt, 240Δ/120-volt
   c) 120/208-volt, 240Δ/208-volt
   d) 277/480-volt, 208Y/120-volt

35) In a building supplied by a feeder(s) two or three single-pole switches or breakers capable of individual operation shall be permitted on multiwire circuits, one pole for each ungrounded conductor, as one multipole disconnect, provided they are equipped with identified handle ties or a master handle to disconnect all ungrounded conductors _____.

   a) So long as each multiwire branch circuit is separately identified
   b) In branch circuits with a nominal voltage of under 600 volts between conductors
   c) With a minimum of 2 grounded conductors supplying a branch circuit fed from the enclosure thereafter
   d) With no more than 6 operations of the hand

36) Which of these are standard ampere ratings for fuses and inverse time circuit breakers?

   a) 15, 20, 60, 75
   b) 80, 90, 350, 110
   c) 20, 25, 115, 155
   d) 300, 400, 550, 1000

37) The operating handle of a circuit breaker shall be permitted to be accessible without opening a door or cover.

   a) True
   b) False

38) Exposed structural metal that is interconnected to form a metal building frame and is not intentionally grounded or bonded and is likely to become energized shall be bonded to which of the following:

   a) Grounded conductor at the service
   b) Disconnecting means for buildings or structures supplied by a feeder or branch circuit
   c) One or more grounding electrodes used, if the grounding electrode conductor or bonding jumper to the grounding electrode is of sufficient size
   d) All of these

39) In overhead services not in excess of 600-volts, nominal, the vertical clearance of final spans above, or within 3 feet measured horizontally of platforms, projections, or surfaces that will permit personal contact shall maintain a minimum clearance of:

   a) 10 feet from final grade above areas or sidewalks accessible only to pedestrians
   b) 12 feet over commercial property and parking areas subject to truck traffic where the voltage does not exceed 300-volts to ground
   c) 15 feet for those areas listed in the 10 feet classification where the voltage exceeds 300-volts to ground
   d) None of these

40) An enclosure designed for surface mounting that has swinging doors or covers secured directly to and telescoping with the walls of the enclosure is:

   a) A panelboard
   b) A switchgear
   c) A cutout box
   d) A cabinet

41) A grounded circuit conductor shall not be used for grounding non-current-carrying metal parts of equipment on the load side of the service disconnecting means or on the load side of a separately derived system disconnecting means or the overcurrent devices for a separately derived system not having a main disconnecting means, except for which of the following:

   a) The frames of ranges, wall mounted ovens, countertop mounted cooking units, and clothes dryers under the conditions permitted for existing installations by 250.140
   b) 125-volt 20-amp receptacles with ground-fault circuit interruption in multifamily dwellings
   c) 125-volt 15-amp isolated ground receptacles in healthcare occupancies
   d) Meter enclosures where service ground-fault protection is installed

42) Luminaires shall be constructed, installed, or equipped with shades or guards so that combustible material is not subjected to temperatures in excess of:

   a) 94° C (201° F)
   b) 75° C (167° F)
   c) 60° C (194°F)
   d) 40° C (104° F)

43) The ampacity of UF cable shall be that of:

   a) 75° C
   b) 60° C
   c) 90° C
   d) 104° C

44) The supply side bonding jumper for a 240-volt single-phase service fed with (2) 300kcmil Aluminum ungrounded conductors should be what size:

   a) 2 copper
   b) 6 copper
   c) 1/0 copper
   d) 4 copper

45) For a one family dwelling with a service rated 400-amps, the service conductors supplying the entire load associated with a one-family dwelling, or the service conductors supplying the entire load associated with an individual dwelling unit in a two-family or multifamily dwelling, shall be permitted to have an ampacity not less than_____of the service rating.

   a) 100%
   b) 125%
   c) 115%
   d) 83%

46) The minimum sized copper equipment grounding conductor required to ground equipment served by a 40-ampere rated branch-circuit is_____.

   a) 14 AWG
   b) 12 AWG
   c) 10 AWG
   d) 8 AWG

47) Receptacles installed in a kitchen to serve countertop surfaces shall be supplied by not fewer than three small-appliance branch circuits.

   a) True
   b) False

48) In a single-family dwelling, the rating of any one cord-and-plug-connected utilization equipment not fastened in place shall not exceed_____ of the branch-circuit ampere rating.

   a) 125%
   b) 100%
   c) 80%
   d) 83%

49) All 15- and 20-ampere, 125- and 250-volt non-locking type receptacles in dwelling units shall be listed tamper-resistant receptacles.

   a) True
   b) False

50) Multiwire branch circuits shall supply only_____.

   a) Line-to-ground loads
   b) Line-to-neutral loads
   c) Line-to-line loads
   d) Three-phase loads

51) Feeder conductors shall be tapped, without overcurrent protection at the tap if:

   a) the ampacity of the tap conductors is not more than the calculated loads on the circuits supplied by the tap conductors
   b) the ampacity of the tap conductors is not more than the rating of the equipment containing an overcurrent protection device(s) supplied by the tap conductors
   c) the tap conductors do not extend beyond the switchboard, switchgear, panelboard, disconnecting means, or control devices they supply
   d) the voltage between any two ungrounded conductors supplying the tap conductors does not exceed 250-volts to ground

52) An incandescent lamp for general use on lighting branch circuits shall not be equipped with a medium base if rating over_____, or with a mogul base if rated over _____.

   a) 150 watts, 1000 watts
   b) 300 watts, 1000 watts
   c) 150 watts, 1500 watts
   d) 300 watts, 1500 watts

53) Type NM, Type NMC, and Type NMS cables shall be permitted to be used in the following, except as prohibited in 334.12:

   a) One- and two-family dwellings and their attached or detached garages, and their storage buildings
   b) Commercial kitchens
   c) Truck Refueling stations
   d) Multi-family dwellings permitted to be of Types I and II construction

54) Type MC cable that provides an effective ground-fault current path in accordance with one or more of the following, can be used as an equipment grounding conductor:

   a) It contains an insulated or uninsulated equipment grounding conductor in compliance with 250.118(1)

   b) the combined metallic sheath and uninsulated equipment grounding/bonding conductor of interlocked metal tape-type MC cable that is listed an identified as an equipment grounding conductor

   c) the metallic sheath or the combined metallic sheath and equipment grounding conductors of the smooth or corrugated tube-type MC cable that is listed and identified as an equipment grounding conductor

   d) all of the above

55) Equipment that utilizes electric energy for electronic, electromechanical, chemical, heating, lighting, or similar purposes is the definition of what?

   a) Cord-and-plug equipment

   b) Equipment grounding conductor

   c) Utilization Equipment

   d) Current Carrying Conductor

56) What is the minimum clearance parameter for an open overhead 240-volt feeder travelling over a pool, from the surface of the water?

   a) 20.5 feet

   b) 22.5 feet

   c) 18 feet

   d) 16 feet

57) For enclosures in wet locations, raceways or cables entering above the level of uninsulated live parts shall use fittings listed for:

   a) Weather-Proof Use
   b) Outdoor use
   c) Damp locations
   d) Wet locations

58) The total rating of utilization equipment fastened in place, other than luminaires, shall _____ of the branch-circuit ampere rating where lighting units, cord-and-plug- connected utilization equipment not fastened in place, or both, are also supplied.

   a) not exceed 50 percent
   b) not exceed 80 percent
   c) be rated at 100 percent
   d) be rated at 125 percent

59) In a separately derived system, a supply-side bonding jumper shall not be required to be larger than:

   a) the derived ungrounded conductors
   b) the grounding electrode conductor
   c) the equipment grounding conductor
   d) the grounded conductor

60) The width of working space in front of electrical equipment shall be the width of the equipment or _____, whichever is greater.

   a) 36 inches
   b) 42 inches
   c) 30 inches
   d) 48 inches

61) All electric pool water heaters shall have the heating elements subdivided into loads not exceeding _____, and protected at not over_____.

   a) 50 amperes, 60 amperes
   b) 48 amperes, 60 amperes
   c) 60 amperes, 80 amperes
   d) 60 amperes, 100 amperes

62) An attachment plug is a:

   a) device designed to open and close a circuit by non-automatic means and to open the circuit automatically on a predetermined overcurrent without damage to itself when properly applied within its rating
   b) device that, by insertion in a receptacle, establishes a connection between the conductors of the attached flexible cord and the conductors connected permanently to the receptacle
   c) device that controls ac/dc voltage or ac/dc current, or both, and that is used to charge a battery or other energy storage device
   d) A fitting intended to terminate a cord to a box or similar device and reduce the strain at points of termination and may include an explosion-proof, a dust-ignition-proof, or a flame-proof seal

63) In damp or wet locations, surface-type meter sockets shall be placed or equipped so as to prevent moisture or water from entering and accumulating within the cabinet or cutout box, and shall be mounted so there is at least_____airspace between the enclosure and the wall or other supporting surface.

   a) 1 inch
   b) ½ inch
   c) 1/8 inch
   d) ¼ inch

64) A concrete-encased electrode shall consist of at least 20 feet of:

   a) bare copper conductor not smaller than 4 AWG
   b) insulated copper conductor not smaller than 4 AWG
   c) bare copper conductor not smaller than 6 AWG
   d) insulated copper conductor not smaller than 6 AWG

65) A 120/208v 3-phase panel with exposed live parts on one side, and no live or grounded parts on the other side of the working space, must have a minimum clear working distance of in front of the panel.

   a) 4 feet
   b) 3 feet 6 inches
   c) 3 feet
   d) 4 feet 6 inches

66) In a trench below 2 inches thick concrete or equivalent, conductors installed in Ridgid Metal Conduit must be buried below a minimum of_____:

   a) 12 inches of covering
   b) 6 inches of covering
   c) 18 inches of covering
   d) 2 inches of covering

67) Underground raceways and cable assemblies entering a hand-hole enclosure shall extend into the enclosure, but they shall not be required to be mechanically connected to the enclosure.

   a) True
   b) False

68) Multi-wire branch circuits shall supply only_____.

   a) Line-to-ground loads
   b) Line-to-neutral loads
   c) Line-to-line loads
   d) Three-phase loads

69) The connection of a grounding electrode conductor or bonding jumper to a grounding electrode shall be made in a manner that will:

   a) ensure an effective grounding path
   b) ensure an effective bonding path
   c) ensure all ungrounded conductors open simultaneously
   d) ensure an effective grounding path

70) Ground-fault circuit-interrupter protection for personnel shall be installed in the branch circuit supplying luminaires operating at: answer:

   a) currents greater than the low-voltage contact limit
   b) voltages lower than the low-voltage contact limit
   c) voltages greater than the low-voltage contact limit
   d) currents lower than the low-voltage contact limit

71) In damp locations (not wet locations) a flush-mounted switch or circuit breaker shall not be required to be equipped with a weather-proof cover.

   a) True
   b) False

72) Feeder conductors shall be tapped, without overcurrent protection at the tap if:

   a) the ampacity of the tap conductors is not more than the calculated loads on the circuits supplied by the tap conductors
   b) the ampacity of the tap conductors is not more than the rating of the equipment containing an overcurrent protection device(s) supplied by the tap conductors
   c) the tap conductors do not extend beyond the switchboard, switchgear, panelboard, disconnecting means, or control devices they supply
   d) the voltage between any two ungrounded conductors supplying the tap conductors does not exceed 250-volts to ground

73) FMC shall not be used in the following:

   a) in dry locations
   b) within 6ft of the outside edge of a water source, such as a bathtub, shower, or sink basin
   c) underground or embedded in poured concrete or aggregate
   d) in dwelling unit attic and storage spaces

74) Ceiling outlets shall be required to support a luminaire weighing a minimum of_____. A luminaire that weighs more than_____ shall be supported independently of the outlet box, unless the outlet box is listed for not less than the weight to be supported.

   a) 45lb, 60lb
   b) 25lb, 50lb
   c) 50lb, 75lb
   d) 50lb, 50lb

75) Screws used for the purpose of attaching receptacles to a box shall be of the type provided with a listed receptacle, or shall be machine screws having_____or part of listed assemblies or systems, in accordance with the manufacturer's instructions.

   a) 18 threads per inch
   b) 21 threads per inch
   c) 30 threads per inch
   d) 32 threads per inch

76) A 4 inch x 2 1/8 inch metal square box, having a minimum volume of 30.3 cubic inches, shall be allowed to have no more than_____12 AWG conductors in it.

   a) 10
   b) 12
   c) 13
   d) 9

77) Splicing of the wire-type grounding electrode conductor shall be permitted only by irreversible compression-type connectors listed as grounding and bonding equipment or by:

   a) the exothermic welding process
   b) the endothermic welding process
   c) listed copper braising
   d) a listed bolt-and-nut termination block

78) A branch-circuit overcurrent protective device is a device capable of providing protection for service, feeder, and branch circuits and equipment over the full range of over-currents between its____ and its_____.

   a) short-circuit, ground-fault
   b) rated over-current rating, voltage rating
   c) rated voltage, HCER rating
   d) rated current, interrupting rating

79) Type SE service-entrance cable shall be permitted for use where the insulated conductors are used for circuit wiring and the uninsulated conductor is used only for equipment grounding purposes – *without considering exceptions.*

a) True
b) False

80) An intersystem bonding termination for connecting intersystem bonding conductors shall be provided_____enclosures at the service equipment or metering equipment enclosures and at the disconnecting means for any additional buildings or structures.

a) internal to
b) external to
c) no closer than 6ft apart near
d) inside all

81) What is the maximum allowed 1/0 AWG THHN conductors can fit in a 1 ½" EMT conduit?

a) 1
b) 3
c) 2
d) 4

82) What is the allowable ampacity for a flexible Type-SO cord with 3 current-carrying conductors?

a) 30-amp
b) 25-amp
c) 20-amp
d) 18-amp

83) Heat-resistant thermoplastic insulation covering 8 AWG conductors are listed for use in _____ and _____ locations?

   a) wet, dry
   b) dry, damp
   c) outdoor, other than dwelling
   d) indoor, other than dwelling

84) At all points where the armor of_____terminates, a fitting shall be provided to protect wires from abrasion, unless the design of the outlet boxes or fittings is such as to afford equal protection, and in addition, an insulating bushing or its equivalent protection shall be provided between the conductors and the armor?

   a) Type MC
   b) Type AC
   c) Type NM
   d) Type UF

85) The total cross-sectional area of 100 feet run of 2 inch EMT conduit is 3.356 square inches, and has 6 12 AWG conductors inside it. What is the total area allowed to be taken up by all conductors in this conduit, regardless of quantity?

   a) 1.342 square inches
   b) 1.566 square inches
   c) 2.343 square inches
   d) 2.013 square inches

86) A single-family dwelling out in the country has a floor area of 1500 square feet, with no garage. Appliances are a 10-kW range and a 6-kW, 240-volt dryer. There are no heating or air conditioning loads, nor a dishwasher or garbage disposal. Using the Standard Method, calculate the minimum size feeder required for this dwelling.

   a) 88-amp
   b) 79-amp
   c) 100-amp
   d) 112-amp

87) When calculating a service load, a load of not less than_____volt-amperes shall be included for each 2-wire laundry branch circuit installed.

   a) 1,200
   b) 950
   c) 1,500
   d) 3,000

88) In a 1-phase 240/120-volt service feeding a Single-Family Dwelling rated 100- through 400- amperes, the service conductors supplying the entire load associated with a one-family dwelling, or the service conductors supplying the entire load associated with an individual dwelling unit in a two-family or multifamily dwelling, shall be permitted to have an ampacity not less than_____.

   a) 100 percent of the service rating
   b) 80 percent of the service rating
   c) 125 percent of the service rating
   d) 83 percent of the service rating

89) The service conductors between the terminals of the service equipment and a point usually outside the building, clear of building walls, where joined by tap or splice to the service drop or overhead service conductors – is defined as what?

   a) Service Entrance Conductors (Overhead)
   b) Service Entrance Conductors (Underground)
   c) Service Equipment
   d) Service Drop

90) An incandescent lamp for general use on lighting branch circuits shall not be equipped with a medium base if rating over_____, or with a mogul base if rated over _____.

   a) 150 watts, 1000 watts
   b) 300 watts, 1000 watts
   c) 150 watts, 1500 watts
   d) 300 watts, 1500 watts

91) A Single-Phase, 3-wire 240-volt service has (2) 2/0 THHN (copper) ungrounded service entrance conductors, what is the minimum size grounding electrode conductor that must be installed?

   a) 4 AWG copper
   b) 6 AWG copper
   c) 1/0 aluminum
   d) 6awg aluminum

92) Copper conductors, for each phase, polarity, neutral, or grounded circuit shall be permitted to be connected in parallel (electrically joined at both ends) only in sizes _____ where installed in other than exempt control circuits.

   a) 1/0 AWG and larger
   b) 2/0 AWG and larger
   c) 1 AWG and larger
   d) 250kcmil and larger

93) A single-family dwelling has a Single-Phase 125-ampere sub-panel in the garage with a main breaker in it rated at 125-amperes. What size equipment grounding conductor shall be used to feed the sub-panel?

   a) 8 AWG copper
   b) 6 AWG copper
   c) 4 AWG copper
   d) 2 AWG copper

94) Conductors that supply one or more welders shall be protected by an overcurrent device rated or set at not more than:

   a) 100 percent of the conductor ampacity
   b) 150 percent of the conductor ampacity
   c) 200 percent of the conductor ampacity
   d) 125 percent of the conductor ampacity

95) In a dwelling unit, receptacles installed within 3 feet of a water heater must be protected by a GFCI receptacle.

   a) True
   b) False

96) Except for equipment specifically listed for operation at 100 percent of its rating, where a branch circuit supplies continuous loads or any combination of continuous and non- continuous loads, the rating of the overcurrent device shall not be less than the non- continuous load plus____of the continuous load.

   a) 115 percent
   b) 83 percent
   c) 200 percent
   d) 125 percent

97) Nonmetallic-sheathed cable shall be supported and secured by staples, or other means, at intervals not exceeding_____and within_____of every cable entry into enclosures such as outlet boxes, junction boxes, cabinets, or fittings.

   a) 3 ½ feet, 12 inches
   b) 3 ½ feet, 30 inches
   c) 4 ½ feet , 12 inches
   d) 4 feet, 12 inches

98) A mobile home floor is 700 square feet and has two small appliance circuits, a 800VA, 240- v heater, 220-VA 120-v exhaust fan, a 400-VA, 120-v dishwasher, and a 6,000-W electric range. What size supply cord is required to supply this mobile home?

   a) 30-amp
   b) 50-amp
   c) 40-amp
   d) 60-amp

99) Which 120-volt, single-phase, 15- and 20-ampere dwelling branch circuits, supplying outlets or devices, shall be protected by an AFCI device?

   a) Family rooms, living rooms, bedrooms
   b) Kitchens, dining rooms, garages
   c) Recreation rooms, closets, exterior patios
   d) Kitchens, libraries, attics

100) Disregarding special conditions or occupancies, a building or other structure that is served by a branch circuit or feeder on the load side of a service disconnecting means shall be supplied by __.

   a) two or less feeders or branch circuits
   b) multiple feeders or branch circuits
   c) only one feeder or branch circuit
   d) none of the above

ANSWERS

100 Practice Questions For The Residential Electrician

Based on the 2017 NFPA 70 National Electrical Code

ANSWER SHEET

1) An insulated grounded conductor of 4 AWG or larger shall be identified by which one of the following means: [200.6 (B)(1)]

   a) A continuous black outer finish
   b) A continuous white outer finish
   c) Three continuous green stripes
   d) None of the above

2) Overhead conductors for festoon lighting shall not be smaller than 12 AWG unless the conductors are: [225.6 (B)]

   a) Conductors that are listed use in damp locations
   b) Of the type THWN, THHN, or XHHW
   c) Supported by messenger wire
   d) No longer than 50 feet in length

3) Where connected to a branch circuit supplying two or more receptacles or outlets, a 30- ampere receptacle shall not supply a total cord-and-plug-connected load in excess of: [210.21 (B)(2) and T210.21 (B)(2)]

   a) 16-amperes
   b) 25-amperes
   c) 30-amperes
   d) 24-amperes

4) A conductor installed on the supply side of a service or within a service equipment enclosure(s), or for a separately derived system, that ensures the required electrical conductivity between metal parts required to be electrically connected is a: [250.2]

   a) Supply-Side Grounding Conductor
   b) Bonding Conductor
   c) Supply-Side Bonding Jumper
   d) Grounding Electrode Conductor

5) In a(n)_____system, electrical equipment, wiring, and other electrically conductive material likely to become energized shall be installed in a manner that creates a low-impedance circuit from any point on the wiring system to the electrical supply source to facilitate the operation of overcurrent devices should a second ground fault from a different phase occur on the wiring system. [250.4 (B)(4)]

   a) Ungrounded
   b) Grounded
   c) 3-phase 4-wire
   d) 1-phase 3-wire

6) The earth shall be considered as an effective fault-current path. [250.4 (B)(4)]

   a) True
   b) False

7) The branch-circuit rating for an appliance that is a continuous load, other than a motor-operated appliance, shall not be less than_____of the marked rating, or not less than _____ of the marked rating if the branch-circuit device and its assembly are listed for continuous loading at 100% of its rating. [422.10(A)]

   a) 125%, 100%
   b) 100%, 125%
   c) 83%, 100%
   d) 100%, 83%

8) Exposed, normally non-current-carrying metal parts of fixed equipment supplied by or enclosing conductors or components that are likely to become energized shall be connected to an equipment grounding conductor under which of the following conditions: [250.110]

   a) Where equipment operates with any terminal at over 150-volts to ground
   b) Where supplied by a wiring method that provides an ungrounded conductor for short sections of metal enclosures
   c) Where 9ft horizontally of ground or grounded metal objects
   d) Where located in an isolated wet or damp location.

9) All pull boxes, junction boxes, and conduit bodies shall be provided with covers compatible with the box or conduit body construction and suitable for the conditions of use. Where used, metal covers shall: [314.28 (C)]

   a) be oversized 3/8 inches to allow for expansion
   b) be used on non-metallic conduit bodies of 2 inches or larger
   c) comply with the grounding requirements of 250.110
   d) be listed for use in wet environments

10) For cord-connected equipment such as room air conditioners, household refrigeration and freezers, drinking water coolers, and beverage dispensers, a separable connector or a(n) _____ shall be permitted to serve as the disconnecting means. [440.12(D)]

    a) none of these
    b) toggle switch
    c) weatherproof cord cap
    d) attachment plug and receptacle

11) Receptacles shall be mounted in identified boxes or assemblies. The boxes or assemblies shall be securely fastened in place unless otherwise permitted elsewhere in this code. Screws for the purpose of attaching receptacles to a box shall be of the type provided with a listed receptacle, or shall be machine screws having_____threads per inch: [406.5]

   a) 18
   b) 32
   c) 21
   d) 20

12) Receptacle outlets in or on floors shall not be counted as part of the required number of receptacle outlets unless located within how many inches of the wall? [210.52(A)(3)]

   a) 6 inches
   b) 12 inches
   c) 18 inches
   d) 24 inches

13) A receptacle outlet shall be installed at each countertop and work surface that is_____ or wider. [210.52(C)(1)]

   a) 6 inches
   b) 12 inches
   c) 18 inches
   d) 36 inches

14) At least_____receptacle(s) outlet shall be installed in bathrooms within_____feet of the outside edge of each basin. [210.52(D)]

   a) 2, 3
   b) 1, 2
   c) 1, 6
   d) 1, 3

15) A single-phase 3-wire 200-amp service is constructed at a residence. What size copper grounding electrode conductor needs to be installed on this service? [T250.66]

   a) #4 copper
   b) #2 copper
   c) #6 copper
   d) #8 copper

16) When installing conductors in a rigid metal conduit that is buried under 3 inches of concrete, to what depth does this conduit need to be buried? [T300.5]

   a) 12 inches
   b) 6 inches
   c) 18 inches
   d) 24 inches

17) Raceways shall be used as a means of support for other raceways under which condition? [300.11(C)]

   a) Where the raceway or means of support is identified as a means of support
   b) Where the raceway is used to support boxes or conduit bodies in accordance with 314.23
   c) Both of these
   d) Neither of these

18) Switches or circuit breakers shall not disconnect the grounded conductor of a circuit. [404.2(B)]

   a) True
   b) False

19) Snap switches rated 20-amperes or less directly connected to aluminum conductors shall be listed and marked_____. [404.14(C)]

   a) ALM/CU
   b) CO/ALR
   c) For use in wet environments
   d) Aluminum conductors only

20) Fixed electric space heating equipment and motors shall be considered what type of load? [424.3(B)]

   a) Inductive
   b) Intermittent
   c) Non-continuous
   d) Continuous

21) In a dwelling unit, which of these locations are NOT required to be protected by and arc- fault circuit interrupter? [210.12(A)]

   a) Bedrooms
   b) Garages
   c) Laundry areas
   d) Hallways

22) GFCI protection must be provided for outlets that supply dishwashers installed in dwelling unit locations. [210.8(D)]

   a) True
   b) False

23) In a bathroom where receptacles are installed within 6 feet from the_____edge of the bowl of a sink, they must be GFCI protected. [210.8(A)(7)]

   a) Bottom outside
   b) Top outside
   c) Point drawn horizontally at the center of the basin
   d) Top inside

24) A 1/0 copper grounding electrode conductor is appropriate for what size ungrounded service entrance conductors? [T250.66]

   a) 2/0 copper
   b) 3/0 copper
   c) Over 350 – 600 copper
   d) Over 3/0 – 350 copper

25) No parts of cord connected luminaires, chain, cable, or cord suspended luminaires, lighting track, pendants, or ceiling suspended (paddle) fans shall be located within a zone measuring__horizontally and_____vertically from the top of the bathtub rim or shower stall threshold. [410.10(D)]

   a) 3 feet, 8 feet
   b) 4 feet, 8 feet
   c) 2 feet, 6 feet
   d) 3 feet, 6 feet

26) Any insulated cable 0-750 volts to ground, supported on and cabled together with a solidly grounded bare messenger or solidly grounded neutral conductor must maintain a clearance of____feet in any direction from the water level, edge of water surface, base of diving platform, or permanently anchored raft. [T680.9(A)]

   a) 6.9
   b) 18
   c) 25
   d) 22.5

27) The ampacity of the branch circuit conductors and the rating or setting of over current protective devices shall not be less than_____of the total nameplate rated load. [680.10]

   a) 100%
   b) 83%
   c) 125%
   d) 115%

28) Temporary electric power and lighting installations shall be permitted for a period not to exceed_____ for holiday decorative lighting and similar purposes. [590.3(B)]

   a) 30 days
   b) 90 days
   c) 60 days
   d) 120 days

29) On a 4-wire, delta connected system where the midpoint of 1 phase winding is grounded, only the conductor or busbar having the higher phase voltage to ground shall be durably and permanently marked by an outer finish that is_____ in color, or by other effective means. [110.15]

   a) Yellow
   b) Orange
   c) Purple
   d) White

30) Exposed live parts on one side of the working space and grounded parts on the other side of the working space must have_____ of clearance in a 277-volt to ground environment. [T110.26(A)(1)]

   a) 3 feet 6 inches
   b) 3 feet
   c) 4 feet
   d) 2 feet 6 inches

31) Which outdoor enclosure types are all approved for wind-blown dust applications? [T110.28]

   a) 3, 3s, 3rx
   b) 3r, 3s, 3rx
   c) 3sx, 3r
   d) 3s, 3x, 3sx

32) For devices with screw shells, the terminal for the_____ conductor shall be connected to the screw shell. [200.10(C)]

   a) Grounded
   b) Ungrounded
   c) Grounding
   d) Equipment grounding

33) A conducting object through which a direct connection to earth is established is termed a(n)___. [100]

   a) Equipment grounding electrode conductor
   b) Grounded conductor
   c) Grounding electrode
   d) Ground bus bar

34) It is appropriate to use the "optional feeder and service load calculation" method to a dwelling unit having the total connected load served by a_____or_____set of 3 wire service or feeder conductors with an ampacity of 100 amps or greater. [220.82]

   a) 120/240-volt, 208Y/120-volt
   b) 120/240-volt, 240Δ/120-volt
   c) 120/208-volt, 240Y/208-volt
   d) 277/480-volt, 208Y/120-volt

35) In a building supplied by a feeder(s) two or three single-pole switches or breakers capable of individual operation shall be permitted on multiwire circuits, one pole for each ungrounded conductor, as one multipole disconnect, provided they are equipped with identified handle ties or a master handle to disconnect all ungrounded conductors _____. [225.33(B)]

   a) So long as each multi-wire branch circuit is separately identified
   b) In branch circuits with a nominal voltage of under 600volts between conductors
   c) With a minimum of 2 grounded conductors supplying a branch circuit fed from the enclosure thereafter
   d) With no more than 6 operations of the hand

36) Which of these are standard ampere ratings for fuses and inverse time circuit breakers? [T240.6(A)]

   a) 15, 20, 60, 75

   b) 80, 90, 350, 110

   c) 20, 25, 115, 155

   d) 300, 400, 550, 1000

37) The operating handle of a circuit breaker shall be permitted to be accessible without opening a door or cover. [240.30(B)]

   a) True

   b) False

38) Exposed structural metal that is interconnected to form a metal building frame and is not intentionally grounded or bonded and is likely to become energized shall be bonded to which of the following: [250.104(C)]

   a) Grounded conductor at the service

   b) Disconnecting means for buildings or structures supplied by a feeder or branch circuit

   c) One or more grounding electrodes used, if the grounding electrode conductor or bonding jumper to the grounding electrode is of sufficient size

   d) All of these

39) In overhead services not in excess of 600 volts, nominal, the vertical clearance of final spans above, or within 3 feet measured horizontally of platforms, projections, or surfaces that will permit personal contact shall maintain a minimum clearance of: [230.24(B)]

   e) 10 feet from final grade above areas or sidewalks accessible only to pedestrians

   f) 12 feet over commercial property and parking areas subject to truck traffic where the voltage does not exceed 300-volts to ground

   g) 15 feet for those areas listed in the 10 feet classification where the voltage exceeds 300-volts to ground

   h) None of these

40) An enclosure designed for surface mounting that has swinging doors or covers secured directly to and telescoping with the walls of the enclosure is: [Article 100]

   a) A panelboard
   b) A switchgear
   c) A cutout box
   d) A cabinet

41) A grounded circuit conductor shall not be used for grounding non-current-carrying metal parts of equipment on the load side of the service disconnecting means or on the load side of a separately derived system disconnecting means or the overcurrent devices for a separately derived system not having a main disconnecting means, except for which of the following: [250.142(B)]

   a) The frames of ranges, wall mounted ovens, countertop mounted cooking units, and clothes dryers under the conditions permitted for existing installations by 250.140
   b) 125-volt 20-amp receptacles with ground-fault circuit interruption in multifamily dwellings
   c) 125-volt 15-amp isolated ground receptacles in healthcare occupancies
   d) Meter enclosures where service ground-fault protection is installed

42) Luminaires shall be constructed, installed, or equipped with shades or guards so that combustible material is not subjected to temperatures in excess of: [410.11]

   a) 94° C (201° F)
   b) 75° C (167° F)
   c) 60° C (194°F)
   d) 40° C (104° F)

43) The ampacity of UF cable shall be that of: [T310.15(B)(16)]

   a) 75° C
   b) 60° C
   c) 90° C
   d) 104° C

44) The supply side bonding jumper for a 240-volt single-phase service fed with (2) 300kcmil Aluminum ungrounded conductors should be what size: [T250.102(C)(1)]

   a) 2 copper
   b) 6 copper
   c) 1/0 copper
   d) 4 copper

45) For a one family dwelling with a service rated 400-amps, the service conductors supplying the entire load associated with a one-family dwelling, or the service conductors supplying the entire load associated with an individual dwelling unit in a two-family or multifamily dwelling, shall be permitted to have an ampacity not less than_____of the service rating. [310.15(B)(7)]

   a) 100%
   b) 125%
   c) 115%
   d) 83%

46) The minimum sized copper equipment grounding conductor required to ground equipment served by a 40-ampere rated branch-circuit is_____. [T250.122]

   a) 14 AWG
   b) 12 AWG
   c) 10 AWG
   d) 8 AWG

47) Receptacles installed in a kitchen to serve countertop surfaces shall be supplied by not fewer than three small-appliance branch circuits. [210.52(B)(3)]

   a) True
   b) False

48) In a single-family dwelling, the rating of any one cord-and-plug-connected utilization equipment not fastened in place shall not exceed_____of the branch-circuit ampere rating. [210.23(A)(1)]

   a) 125%
   b) 100%
   c) 80%
   d) 83%

49) All 15- and 20-ampere, 125- and 250-volt non-locking type receptacles in dwelling units shall be listed tamper-resistant receptacles. [406.12]

   a) True
   b) False

50) Multi-wire branch circuits shall supply only_____. [210.4(c)]

   a) Line-to-ground loads
   b) Line-to-neutral loads
   c) Line-to-line loads
   d) Three-phase loads

51) Feeder conductors shall be tapped, without overcurrent protection at the tap if: [240.21(B)(1)(2)]

   a) the ampacity of the tap conductors is not more than the calculated loads on the circuits supplied by the tap conductors

   b) the ampacity of the tap conductors is not more than the rating of the equipment containing an overcurrent protection device(s) supplied by the tap conductors

   c) the tap conductors do not extend beyond the switchboard, switchgear, panelboard, disconnecting means, or control devices they supply

   d) the voltage between any two ungrounded conductors supplying the tap conductors does not exceed 250-volts to ground

52) An incandescent lamp for general use on lighting branch circuits shall not be equipped with a medium base if rating over_____, or with a mogul base if rated over _____. [410.103]

   a) 150 watts, 1000 watts

   b) 300 watts, 1000 watts

   c) 150 watts, 1500 watts

   d) 300 watts, 1500 watts

53) Type NM, Type NMC, and Type NMS cables shall be permitted to be used in the following, except as prohibited in 334.12: [334.10]

   a) One- and two-family dwellings and their attached or detached garages, and their storage buildings

   b) Commercial kitchens

   c) Truck Refueling stations

   d) Multi-family dwellings permitted to be of Types I and II construction

54) Type MC cable that provides an effective ground-fault current path in accordance with one or more of the following, can be used as an equipment grounding conductor: [250.118(10)]

   a) It contains an insulated or uninsulated equipment grounding conductor in compliance with 250.118(1)

   b) the combined metallic sheath and uninsulated equipment grounding/bonding conductor of interlocked metal tape-type MC cable that is listed an identified as an equipment grounding conductor

   c) the metallic sheath or the combined metallic sheath and equipment grounding conductors of the smooth or corrugated tube-type MC cable that is listed and identified as an equipment grounding conductor

   d) all of the above

55) Equipment that utilizes electric energy for electronic, electromechanical, chemical, heating, lighting, or similar purposes is the definition of what? [Article 100]

   a) Cord-and-plug equipement
   b) Equipment grounding conductor
   c) Utilization Equipment
   d) Current Carrying Conductor

56) What is the minimum clearance parameter for an open overhead 240-volt feeder travelling over a pool, from the surface of the water? [T680.9(A)]

   a) 20.5 feet
   b) 22.5 feet
   c) 18 feet
   d) 16 feet

57) For enclosures in wet locations, raceways or cables entering above the level of uninsulated live parts shall use fittings listed for: [312.2]

   a) Weather-Proof Use
   b) Outdoor use
   c) Damp locations
   d) Wet locations

58) The total rating of utilization equipment fastened in place, other than luminaires, shall _____ of the branch-circuit ampere rating where lighting units, cord-and-plug- connected utilization equipment not fastened in place, or both, are also supplied. [210.23(A)(2)]

   a) not exceed 50 percent
   b) not exceed 80 percent
   c) be rated at 100 percent
   d) be rated at 125 percent

59) In a separately derived system, a supply-side bonding jumper shall not be required to be larger than: [250.30(A)(2)]

   a) the derived ungrounded conductors
   b) the grounding electrode conductor
   c) the equipment grounding conductor
   d) the grounded conductor

60) The width of working space in front of electrical equipment shall be the width of the equipment or _____, whichever is greater. [110.26(A)(2)]

   a) 36 inches
   b) 42 inches
   c) 30 inches
   d) 48 inches

61) All electric pool water heaters shall have the heating elements subdivided into loads not exceeding _____, and protected at not over_____. [680.10]

   a) 50 amperes, 60 amperes

   b) 48 amperes, 60 amperes

   c) 60 amperes, 80 amperes

   d) 60 amperes, 100 amperes

62) An attachment plug is a: [Article 100]

   a) device designed to open and close a circuit by nonautomatic means and to open the circuit automatically on a predetermined overcurrent without damage to itself when properly appied within its rating

   b) device that, by insertion in a receptacle, establishes a connection between the conductors of the attached flexible cord and the conductors connected permanently to the receptacle

   c) device that controls ac/dc voltage or ac/dc current, or both, and that is used to charge a battery or other energy storage device

   d) A fitting intended to terminate a cord to a box or similar device and reduce the strain at points of termination and may include an explosion-proof, a dust-ignition-proof, or a flame-proof seal

63) In damp or wet locations, surface-type meter sockets shall be placed or equipped so as to prevent moisture or water from entering and accumulating within the cabinet or cutout box, and shall be mounted so there is at least ¼" in airspace between the enclosure and the wall or other supporting surface. [312.2]

   a) 1 inch

   b) ½ inch

   c) 1/8 inch

   d) ¼ inch

64) A concrete-encased electrode shall consist of at least 20 feet of: [250.52(A)(3)]

   a) bare copper conductor not smaller than 4 AWG
   b) insulated copper conductor not smaller than 4 AWG
   c) bare copper conductor not smaller than 6 AWG
   d) insulated copper conductor not smaller than 6 AWG

65) A 120/208v 3-phase panel with exposed live parts on one side, and no live or grounded parts on the other side of the working space, must have a minimum clear working distance of_in front of the panel. [T110.26(A)(1)]

   a) 4 feet
   b) 3 feet 6 inches
   c) 3 feet
   d) 4 feet 6 inches

66) In a trench below 2 in thick concrete or equivalent, conductors installed in Ridgid Metal Conduit must be buried below a minimum of_____: [T300.5]

   a) 12 inches of covering
   b) 6 inches of covering
   c) 18 inches of covering
   d) 2 inches of covering

67) Underground raceways and cable assemblies entering a hand-hole enclosure shall extend into the enclosure, but they shall not be required to be mechanically connected to the enclosure. [314.30(B)]

   a) True
   b) False

68) Multi-wire branch circuits shall supply only_____. [210.4(c)]

   a) Line-to-ground loads
   b) Line-to-neutral loads
   c) Line-to-line loads
   d) Three-phase loads

69) The connection of a grounding electrode conductor or bonding jumper to a grounding electrode shall be made in a manner that will: [250.68(B)]

   a) ensure an effective grounding path
   b) ensure an effective bonding path
   c) ensure all ungrounded conductors open simultaneously
   d) ensure an effective grounding path

70) Ground-fault circuit-interrupter protection for personnel shall be installed in the branch circuit supplying luminaires operating at: answer: [680.23 (A)(3)]

   a) currents greater than the low-voltage contact limit
   b) voltages lower than the low-voltage contact limit
   c) voltages greater than the low-voltage contact limit
   d) currents lower than the low-voltage contact limit

71) In damp locations (not wet locations) a flush-mounted switch or circuit breaker shall not be required to be equipped with a weather-proof cover. [404.4(B)]

   a) True
   b) False

72) Feeder conductors shall be tapped, without overcurrent protection at the tap if: [240.21(B)(1)(2)]

   a) the ampacity of the tap conductors is not more than the calculated loads on the circuits supplied by the tap conductors
   b) the ampacity of the tap conductors is not more than the rating of the equipment containing an overcurrent protection device(s) supplied by the tap conductors
   c) the tap conductors do not extend beyond the switchboard, switchgear, panelboard, disconnecting means, or control devices they supply
   d) the voltage between any two ungrounded conductors supplying the tap conductors does not exceed 250-volts to ground

73) FMC shall not be used in the following: [348.12(6)]

   a) in dry locations
   b) within 6ft of the outside edge of a water source, such as a bathtub, shower, or sink basin
   c) underground or embedded in poured concrete or aggregate
   d) in dwelling unit attic and storage spaces

74) Ceiling outlets shall be required to support a luminaire weighing a minimum of_____. A luminare that weighs more than_____shall be supported independently of the outlet box, unless the outlet box is listed for not less than the weight to be supported. [314.27(A)(2)]

   a) 45lb, 60lb
   b) 25lb, 50lb
   c) 50lb, 75lb
   d) 50lb, 50lb

75) Screws used for the purpose of attaching receptacles to a box shall be of the type provided with a listed receptacle, or shall be machine screws having_____or part of listed assemblies or systems, in accordance with the manufacturer's instructions. [406.5]

   a) 18 threads per inch
   b) 21 threads per inch
   c) 30 threads per inch
   d) 32 threads per inch

76) A 100x54mm (4x2 1/8") metal square box, having a minimum volume of 30.3 cubic inches, shall be allowed to have no more than_____12 AWG conductors in it. [T314.16 (A)]

   a) 10
   b) 12
   c) 13
   d) 9

77) Splicing of the wire-type grounding electrode conductor shall be permitted only by irreversible compression-type connectors listed as grounding and bonding equipment or by: [250.64(C)(1)]

   a) the exothermic welding process
   b) the endothermic welding process
   c) listed copper braising
   d) a listed bolt-and-nut termination block

78) A branch-circuit overcurrent protective device is a device capable of providing protection for service, feeder, and branch circuits and equipment over the full range of over-currents between its____ and its_____. [Article 100]

   a) short-circuit, ground-fault
   b) rated over-current rating, voltage rating
   c) rated voltage, HCER rating
   d) rated current, interrupting rating

79) Type SE service-entrance cable shall be permitted for use where the insulated conductors are used for circuit wiring and the uninsulated conductor is used only for equipment grounding purposes – *without considering exceptions.* [338.10(B)(2)]

   a) True
   b) False

80) An intersystem bonding termination for connecting intersystem bonding conductors shall be provided_____enclosures at the service equipment or metering equipment enclosures and at the disconnecting means for any additional buildings or structures. [250.94(A)]

   a) internal to
   b) external to
   c) no closer than 6ft apart near
   d) inside all

81) What is the maximum allowed 1/0 AWG THHN conductors can fit in a 1 ½" EMT conduit? [Ch. 9 Table 1]

   a) 1
   b) 3
   c) 2
   d) 4

82) What is the allowable ampacity for a flexible Type-SO cord with 3 current-carrying conductors? [T400.5(A)(1)]

   a) 30-amp
   b) 25-amp
   c) 20-amp
   d) 18-amp

83) Heat-resistant thermoplastic insulation covering 8 AWG conductors are listed for use in _____ and _____ locations? [T310.104(A)]

   a) wet, dry
   b) dry, damp
   c) outdoor, other than dwelling
   d) indoor, other than dwelling

84) At all points where the armor of _____ terminates, a fitting shall be provided to protect wires from abrasion, unless the design of the outlet boxes or fittings is such as to afford equal protection, and in addition, an insulating bushing or its equivalent protection shall be provided between the conductors and the armor? [320.40]

   a) Type MC
   b) Type AC
   c) Type NM
   d) Type UF

85) The total cross-sectional area of 100 feet run of 2 inch EMT conduit is 3.356 square inches, and has 6 12 AWG conductors inside it. What is the total area allowed to be taken up by all conductors in this conduit, regardless of quantity? [Ch. 9 Table 4]

   a) 1.342 square inches
   b) 1.566 square inches
   c) 2.343 square inches
   d) 2.013 square inches

86) A single-family dwelling out in the country has a floor area of 1500 square feet, with no garage. Appliances are a 10-kW range and a 6-kW, 240-volt dryer. There are no heating or air conditioning loads, nor a dishwasher or garbage disposal. Using the Standard Method, calculate the minimum size feeder required for this dwelling. [220.40 – various sections]

   a) 88-amp

   b) 79-amp

   c) 100-amp

   d) 112-amp

87) When calculating a service load, a load of not less than _____ volt-amperes shall be included for each 2-wire laundry branch circuit installed. [220.52(B)]

   a) 1,200

   b) 950

   c) 1,500

   d) 3,000

88) In a 1-phase 240/120-volt service feeding a Single-Family Dwelling rated 100- through 400- amperes, the service conductors supplying the entire load associated with a one-family dwelling, or the service conductors supplying the entire load associated with an individual dwelling unit in a two-family or multifamily dwelling, shall be permitted to have an ampacity not less than _____. [310.15(B)(7)]

   a) 100 percent of the service rating

   b) 80 percent of the service rating

   c) 125 percent of the service rating

   d) 83 percent of the service rating

89) The service conductors between the terminals of the service equipment and a point usually outside the building, clear of building walls, where joined by tap or splice to the service drop or overhead service conductors – is defined as what? [Article 100]

   a) Service Entrance Conductors (Overhead)
   b) Service Entrance Conductors (Underground)
   c) Service Equipment
   d) Service Drop

90) An incandescent lamp for general use on lighting branch circuits shall not be equipped with a medium base if rating over_____, or with a mogul base if rated over _____. [410.103]

   a) 150 watts, 1000 watts
   b) 300 watts, 1000 watts
   c) 150 watts, 1500 watts
   d) 300 watts, 1500 watts

91) A Single-Phase, 3-wire 240-volt service has (2) 2/0 THHN (copper) ungrounded service entrance conductors, what is the minimum size grounding electrode conductor that must be installed? [T250.66]

   a) 4 AWG copper
   b) 6 AWG copper
   c) 1/0 aluminum
   d) 6awg aluminum

92) Copper conductors, for each phase, polarity, neutral, or grounded circuit shall be permitted to be connected in parallel (electrically joined at both ends) only in sizes _____ where installed in other than exempt control circuits. [310.10(H)(1)]

   a) 1/0 AWG and larger
   b) 2/0 AWG and larger
   c) 1 AWG and larger
   d) 250kcmil and larger

93) A single-family dwelling has a Single-Phase 125-ampere sub-panel in the garage with a main breaker in it rated at 125-amperes. What size equipment grounding conductor shall be used to feed the sub-panel? [T250.155]

   a) 8 AWG copper
   b) 6 AWG copper
   c) 4 AWG copper
   d) 2 AWG copper

94) Conductors that supply one or more welders shall be protected by an overcurrent device rated or set at not more than: [630.12(B)]

   a) 100 percent of the conductor ampacity
   b) 150 percent of the conductor ampacity
   c) 200 percent of the conductor ampacity
   d) 125 percent of the conductor ampacity

95) In a dwelling unit, receptacles installed within 3 feet of a water heater must be protected by a GFCI receptacle. [210.8(A)]

   a) True
   b) False

96) Except for equipment specifically listed for operation at 100 percent of its rating, where a branch circuit supplies continuous loads or any combination of continuous and non- continuous loads, the rating of the overcurrent device shall not be less than the non- continuous load plus____of the continuous load. [210.20(A)]

   a) 115 percent

   b) 83 percent

   c) 200 percent

   d) 125 percent

97) Nonmetallic-sheathed cable shall be supported and secured by staples, or other means, at intervals not exceeding_____and within_____of every cable entry into enclosures such as outlet boxes, junction boxes, cabinets, or fittings. [334.30]

   a) 3 ½ feet, 12 inches

   b) 3 ½ feet, 30 inches

   c) 4 ½ feet , 12 inches

   d) 4 feet, 12 inches

98) A mobile home floor is 700 sq-ft and has two small appliance circuits, a 800-VA, 240-v heater, 220-VA 120-v exhaust fan, a 400-VA, 120-v dishwasher, and a 6,000-W electric range. What size supply cord is required to supply this mobile home? [550.18]

   a) 30-amp

   b) 50-amp

   c) 40-amp

   d) 60-amp

99) Which 120-volt, single-phase, 15- and 20-ampere dwelling branch circuits, supplying outlets or devices, shall be protected by an AFCI device? [210.12(A)]

   a) Family rooms, living rooms, bedrooms
   b) Kitchens, dining rooms, garages
   c) Recreation rooms, closets, exterior patios
   d) Kitchens, libraries, attics

100) Disregarding special conditions or occupancies, a building or other structure that is served by a branch circuit or feeder on the load side of a service disconnecting means shall be supplied by __. [225.30]

   a) two or less feeders or branch circuits
   b) Multiple feeders or branch circuits
   c) only one feeder or branch circuit
   d) none of the above

If you haven't already done so, join the Electrical Wizardry Facebook group and check out the Electrician U YouTube Channel, Journey 2 Master YouTube Channel, and the Journey 2 Master podcast on iTunes and Spotify. There are a ton of good videos on there from in the field that will help you understand NON-CODE things. There's also many tool reviews and discussions that may help you get your head wrapped around bigger concepts and procedures for how to do your job better. Or at least maybe just how to understand things in a way you haven't thought of them before. Please get involved in the electrical community, ask questions, and share your knowledge with others coming up behind you. It will make you a better teacher and will help you understand this trade in new and interesting ways you didn't realize were possible.

Good luck on the test my friend, and let me know if you passed!